Aliens are Already Among Us

Copyright Page

Aliens are Already Among Us

The Aliens and UFO Secrets Series Book 2

By Martin K. Ettington

All Rights Reserved USA 2019

Printed in the United States of America

ISBN: 9781098629946

Aliens are Already Among Us

Aliens are Already Among Us

Other books by Martin K. Ettington

Spiritual and Metaphysics Books:
Prophecy: A History and How to Guide
God Like Powers and Abilities
Enlightenment for Newbies
Removing Illusions to Find True Happiness
Using the Scientific Method to Study the Paranormal
A Compendium of Metaphysics and How to Guides (Six books together in one volume)
Love from the Heart
The Enlightenment Experience
Learn Your Soul's Purpose
Pursuing Enlightenment
A Modern Man's Search for Truth
Use Intuition and Prophecy to Improve Your Life
The Handbook of Spiritual and Energy Healing

Longevity & Immortality:
Physical Immortality: A History and How to Guide
The Commentaries of Living Immortals
Records of Extremely Long Lived Persons
Enlightenment and Immortality
Longevity Improvements from Science
The 10 Principles of Personal Longevity
Telomeres & Longevity
The Diets and Lifestyles of the Worlds Oldest Peoples
The Longevity Six Books Bundle

Science Fiction:
Out of This Universe
Personal Freedom-Parts 1 & 2
The Psychic Soldier Series:
  Book 1-Himalayan Journey
  Book 2-A Soldier is Born
  Book 3-Fighting For Right
  Book 4-Earth Protector
The Immortality Sci Fi Bundle

The God Like Powers Series:
Human Invisibility
Invulnerability and Shielding
Teleportation
Psychokinesis
Our Energy Body, Auras, and Thoughtforms

The God Like Powers Series—Volume 1 Compilation

The Yoga Discovery Series:
Yoga-An Ancient Art Form
Hatha Yoga-Helping you Live Better
Raja Yoga-Through the Ages
The Yoga Discovery Package

Business & Coaching Books:
Creating, Publishing, & Marketing Practitioner Ebooks
Building a Successful Longevity Coaching Business
Why Become a Coach?
The Professional Coaching Success Trilogy
2020-Make Money Writing and Selling Books
The 2020 Handbook of High Paying Work Without a College Degree

Science, Technology, and Misc.
Future Predictions By and Engineer & Seer
The Unusual Science & Technology Bundle
The Real Atlantis-In the Eye of the Sahara
Are Cryptozoological Animals Real or Imaginary?
Real Time Travel Stories From a Psychic Engineer
Removing Limits On Our Consciousness-And Thinking Outside the Box
33 Incredible True Survival Stories
How to Survive Anything: From the Wilderness to Man Made Disasters
All About Mars Journeys and Settlement
Mining the Asteroid Belt

Ancient History
The Real Atlantis-In the Eye of the Sahara
Ancient & Prehistoric Civilizations
Ancient & Prehistoric Civilizations-Book Two
The History of Antediluvian Giants
The Antediluvian History of Earth
Ancient Underground Cities and Tunnels
Strange Objects Which Should Not Exist
Strange and Ancient Places in the USA
A Theory of Ancient Prehistory And Giant Aliens

Aliens and Space
Aliens and Secret Technology
Aliens Are Already Among Us
Designing and Building Space Colonies
Humanity and the Universe
All About Moon Bases
All About Mars Journeys and Settlement

Aliens are Already Among Us

The Space and Aliens Six Books Bundle
A Theory of Ancient Prehistory and Giant Aliens

The Space Colonies and Space Structures Coloring Book
All About Asteroids

<u>The Longevity Training Series</u>

(A transcription of the online Multimedia Longevity Coaching Training Program)

The Personal Longevity Training Series-Book1-Long Lived Persons
The Personal Longevity Training Series-Book2-Your Soul's Purpose
The Personal Longevity Training Series-Book3-Enable Your Life Urge
The Personal Longevity Training Series-Book4-Your Spiritual Connection
The Personal Longevity Training Series-Book5-Having Love in Your Heart
The Personal Longevity Training Series-Book6-Energy Body Health
The Personal Longevity Training Series-Book7-The Science of Longevity
The Personal Longevity Training Series-Book8-Physical Body Health
The Personal Longevity Training Series-Book9-Avoiding Accidents
The Personal Longevity Training Series-Book10-Implementing These Principles

The Personal Longevity Training Series-Books One Thru Ten

These books are all available in digital and printed formats from my website and on Amazon, Barnes & Noble, Apple ITunes, and many other sites

My Books Website is: http://mkettingtonbooks.com

Aliens are Already Among Us

The focus of this book is on Aliens having visited or living on this Earth. There is lots of evidence that not only are Aliens on Earth during the present, but have been here for thousands if not millions of years.

There are lots of stories and some good evidence that this is all true. Even today there are videos of alien craft, and stories in the news about our government still studying and reporting on Aliens visiting us.

- What is this evidence of Aliens living on the Earth in the past and currently?
- And where do we separate the facts from the fiction?
- What technologies might these Aliens have?
- And how do these beings interact with our civilization?

Lots of information is provided here to help better understand who the aliens are, how many types we know of, and how it might be possible for them to stay hidden on the Earth for thousands of years.

Hope your find all this information interesting and it provides you with some food for thought.

Aliens are Already Among Us

Aliens are Already Among Us

Signup for our Mailing List to get the following:

1) A discount coupon for 25% discount on all books on our site

2) Occasional Notices of new books available

3) Occasional Email on other offerings of ours (Monthly)

Go to this link to sign-up:

http://personal-longevity.com/mkebooks/emailsignup/

And click this link to get the FREE 102 page Ebook titled "Secrets of Many Things"

If you have any questions about this book or other subjects please contact the Author at:

mke@mkettingtonbooks.com

Aliens are Already Among Us

Aliens are Already Among Us

Table of Contents

1.0 Introduction 1
2.0 The Probability of Aliens Existing 3
   a. The Drake equation 3
   b. What Astronomy and Astrophysics tells us 4
   c. The possibility of Advanced Alien Life 5
3.0 The History of UFO & Aliens 9
   a. Ancient grooved spheres 9
   b.   Legends of the Annunaki 10
   c. Paintings on Stone walls in Northern Italy 12
   d. The story of Ezekiel and the flying machine 13
   e. Details of a fresco entitled "The Crucifixion" 14
   f. The Dogon Tribe in Africa 15
   g. UFO Sightings by the Puritans 17
   h. Sightings in the Nineteenth Century 21
4.0 Roswell and EBEs 25
5.0 Serpo—Missions to Other Planets 31
6.0 Alien Bases on Earth 37
   a.   Indian Springs, California 37
   b.   The Dulce Base 40
   c.   Additional Alien Underground Bases: 43
   Underground alien bases in Europe 43
   Underground alien bases in Canada 44
   Underground alien bases in Russia 46
   Underground alien bases in America 47
7.0 Aliens Underwater 53
   a.   Underwater Base off The Coast Of Malibu 53
   b.   Underwater Base Off Puffin Island Wales 55
   c.   UFO Base Under Lake Ontario 56
   d.   Flight Corridor to Underwater Base North Island, New Zealand 58
   e.   Pacific Ocean 'Humming' Is Actually Marine Life Breaking Wind 59
   f.   Strange Lights and Sounds in Puerto Rico 60
   g.   Ancient Alien Base at Lake Titicaca 62
   h.   Underwater Alien Base at Guantanamo Bay 63
   i.   Increased UFO Sightings At Lake Erie Cleveland, Ohio 65

Aliens are Already Among Us

    j.   Lake Baikal—Scene of an Underwater Battle with Aliens 66

8.0 NATO's Top Secret Study 69
9.0 Other Interesting Information 77
    Aliens and Volcanos 78
10.0 Alien Technology 81
    a.   Alien Invisibility 81
    b.   Caret Alien Technology: 81
11.0 Aliens on the Astral Plane 87
12.0 How to contact Aliens 91
13.0 Summary 99
Bibliography 101
Index 103

Aliens are Already Among Us

1.0 Introduction

This book is all about Aliens having visited or are living on this Earth in the past and today. There is lots of evidence that not only are Aliens on Earth during the present, but have been here for thousands if not millions of years.

There are lots of stories and some good evidence that this is all true. Even today there are videos of alien craft, and stories in the news about our government still studying and reporting on Aliens visiting us.

Here is a recent picture of a military jet showing an alien craft in flight in 2004:

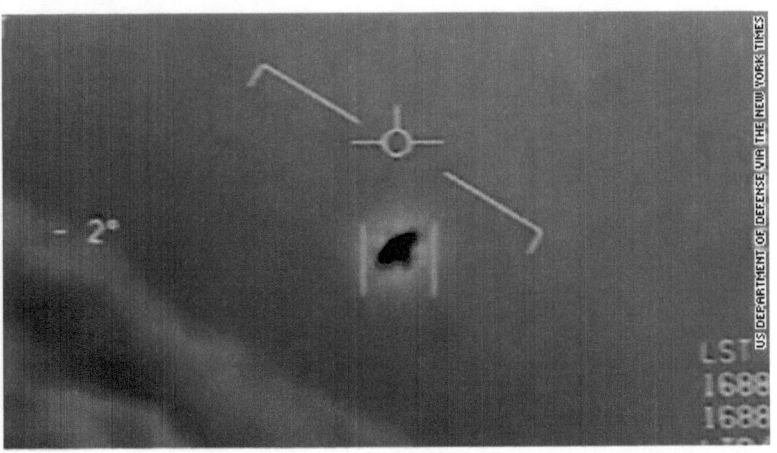

What is this evidence of Aliens living on the Earth in the past and currently?

And where do we separate the facts from the fiction?

What technologies might these Aliens have?

And why are these beings interested in our civilization?

Aliens are Already Among Us

Lots of information is provided here to help better understand who the aliens are, how many types we know of, and how it might be possible for them to stay hidden on the Earth for thousands of years.

Hope you find all this information interesting and it provides you with some food for thought.

2.0 The Probability of Aliens Existing

In this chapter we explore what science and astronomy can tell us about the likelihood of alien races in our galaxy. (I. E. The Milky Way)

a. The Drake equation

Dr. Frank Drake an astronomer proposed an equation on the possibility of intelligent life at the first scientific meeting on the search for extraterrestrial life in 1960. (SETI). This equation has since become the definitive calculation for the existence of alien life.

The Drake equation is shown below with definitions of its variables.

$$N = R^* \, f_p \, n_e \, f_l \, f_i \, f_c \, L$$

- N = The number of communicative civilizations
- R^* = The rate of formation of suitable stars (stars such as our Sun)
- f_p = The fraction of those stars with planets. (Current evidence indicates that planetary systems may be common for stars like the Sun.)
- n_e = The number of Earth-like worlds per planetary system
- f_l = The fraction of those Earth-like planets where life actually develops
- f_i = The fraction of life sites where intelligence develops
- f_c = The fraction of communicative planets (those on which electromagnetic communications technology develops)
- L = The "lifetime" of communicating civilizations

Once you fill in the variables this equation will tell you an estimate of the number of civilizations on other planets that we should be able to communicate.

Now that we are finding lots of planets and expect to find many planets that can support life, the probability of intelligent life has gone up significantly.

When filling in the variables based on conservative assumptions and our current knowledge of others stars, we

Aliens are Already Among Us

find that there should be thousands of intelligent civilizations in just our own galaxy to talk to.

b. What Astronomy and Astrophysics tells us

A good place to start is with what our current understanding of science and astronomical observations of the Universe tell us is possible.

Most Astronomers calculate that the age of our Universe is between 13.5 to 14 billion years old. That there are also galaxies and stars that were formed only a billion years after the big bang that started our universe.

Life on earth took several billion years to start and our planet is estimated to be 4.5 billion years old. Therefore, if life exists on any planets in the universe it may have started as much as nine to ten billion years ago.

In the 1990s our telescopes and astronomy tools also became good enough to start finding planets around other stars. As of 2017, 4,500 candidate exoplanets have been found around stars by the Kepler Observatory. Of these about ten are Earth like planets in that they are in the stars habitable zone and are rocky planets like Earth. Almost every star examined seems to have planets.

We also know that there are at least 100 billion stars in our galaxy, and maybe more. (There are an estimated two trillion galaxies overall.) Let's also make an assumption that there might be one out of one thousand Earth like worlds which have intelligent life. Given the above information collected from Kepler and other observatories we can conclude the following:

Aliens are Already Among Us

That there should be over 100 million planets in our Galaxy alone which support intelligent life. That is a lot of potential alien civilizations.

c. The possibility of Advanced Alien Life

If life existed or exists that started in the first galaxies then what would it be like today?

We may speculate that any life that old would have evolved to be as Gods to us or as advanced compared to us as we are to an Amoeba.

Even life on other planets that evolved in a similar timeframe to ours could be thousands or even millions of years more advanced.

What types of technologies and abilities would beings even thousands of years older than us possess?

Just looking at the advancement of technology over time we should expect Aliens to be able to do things which would seem as magic to us.

Take stealth aircraft for example: Today scientists are already experimenting on technologies to bend light around materials to provide full invisibility. This should be attainable in a few more years.

Of course I'm also assuming that very advanced Aliens would be able to find a way around the relativistic light speed barrier to travel easily between stars.

Aliens are Already Among Us

In terms of their ability to hide themselves, they would have stealth capabilities for their ships and individuals that would include the ability to hide from us completely if they so desired.

Here is a picture taken in Chile as part of a video which the photographer says is undoctored and might be an alien who thought he was invisible:

I've studied and experienced a lot of paranormal events in my life so I'm a strong believer in intelligent beings having these abilities too. In terms of mental and paranormal abilities I would also speculate that Aliens would be well advanced beyond ninety nine percent of humans alive today too.

If Aliens are so far advanced of us mentally and in paranormal abilities they might also be able to make us see illusions and control us with their minds if they were here on earth. The possibilities are endless.

The main reason to discuss all these possibilities is to get the reader to understand that true Alien life may be so far

Aliens are Already Among Us

beyond us that we could not even comprehend their abilities.

Therefore, we shouldn't summarily dismiss even the craziest reports from people who report UFO and Alien experiences without a proper investigation.

Here is another picture from the 1950s of an alleged meeting between the US Military and an Alien. (Who looks like a small grey):

An old man claims to have worked with the Aliens at Area 51 over fifty years ago:

3.0 The History of UFOs & Aliens

There are many historical records that UFOs and Aliens have been observed in recorded history. There is also lots of evidence of anomalous technology that humans couldn't have created. Many of these anomalies existed before humans could have evolved and lived on Earth. Some of the best known include:

a. Ancient grooved spheres

Over the last few decades, miners in South Africa have been digging up mysterious metal spheres. Origin unknown, these spheres measure approximately an inch or so in diameter, and some are etched with three parallel grooves running around the equator. Two types of spheres have been found: one is composed of a solid bluish metal with flecks of white; the other is hollowed out and filled with a spongy white substance. The kicker is that the rock in which they were found is Precambrian - and dated to 2.8 *billion* years old!

Who made them and for what purpose is unknown. Something this old would have to have been created before any life existed on earth.

b. Legends of the Annunaki

There are lots books written on the subject of the Anunnaki who many of the Authors think were extraterrestrials meeting and affecting human civilization thousands of years ago.

Originally, the Anunnaki appear to have been heavenly deities with immense powers. In *Enki and the World Order*, the Anunnaki "do homage" to Enki, sing hymns of praise in his honor, and "take up their dwellings" among the people of Sumer. The same composition repeatedly states that the Anunnaki "decree the fates of mankind"

The ancient Sumerians claimed they were aliens who had ruled Earth for many centuries. Here is a short summary what the Sumerians believed:

> *The Sumerians possessed advanced understanding of mathematics, language and astronomy. However, there is no evidence of this knowledge evolving over time as one would expect. It just exists and is written about as being brought by the Anuna gods who came from the stars. The*

Aliens are Already Among Us

Sumerians are credited with establishing our modern day calendar and knew the cycles of the sun, moon, and visible planets. They knew the accurate length of a year.

The Sumerians were the first civilization to divide space and time by factors of six.

The contemporary division of the year into 12 months, 24 hours days, hours into 60 minutes with 60 seconds are all documented by the Sumerians. This division by factors of six was later found at several other megalithic structures on other continents.

Sumerian understanding of time gives perspective to the writings known as the "<u>Kings List</u>." This chronicle of rule begins with original Anuna supreme rulers' decent from the heavens in the "kingship". This is said to have happened after a great flood. This is believed to be referring to the Biblical flood. It then lists all the Anuna rulers that followed and how long they ruled. Amazingly, many of them are listed to rule for hundreds of years.

Other for much shorter times (30 years on average.) The first ruler held power for 1200 years! We cannot imagine this to be true since our human life span is 70-90 years on average. However, consider that a highly advanced being, millions of years more developed than we are, may have achieved this kind of longevity. Consider our own life span average of 70-90 years compared to that of humans during the first century BC that lived 30 years on average. Now, imagine what effect a million years of evolution may have on our life span.

The Kings List chronicles 433,000 years of Anuna rule. This means the age of the Anuna gods spanned a much longer period of time than the recorded history of all civilizations on Earth combined. The Anuna/Annunaki would have been on earth 430,000 years before any evidence of recorded history that has been discovered.

We cannot imagine this amount of time since our limited experience in recorded history is less than 6,000 years in total (from Mesopotamia to today.) The accounts of people living these lengths of time are also found in the Hebrew Torah and Christian Bible's texts called the Old Testament. The real question is just how old are these accounts of time? Likely much older than any of us can possibly imagine.

c. Paintings on Stone walls in Northern Italy

Paintings on stonewalls in Val Camonica, North Italy, showing beings seemingly wearing a glasslike helmet from

Aliens are Already Among Us

which short rays are emitted. Dated 4,000 B.C. These could be some beings in spacesuits

d. The story of Ezekiel and the flying machine

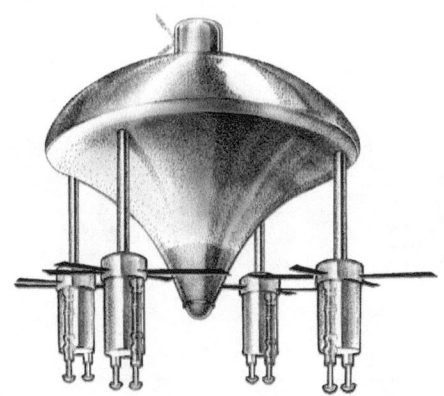

In about 600 B.C Ezekiel made one of the earliest accounts of some type of flying machine. It is recorded in the Old Testament of the Bible and here are some of the passages:

> *And I looked, and, behold, a whirlwind came out of the north, a great cloud, and a fire infolding itself, and a brightness was about it, and out of the midst thereof as the color of amber, out of the midst of the fire. (Ezek. 1:4)*

> *And I looked, and, behold, a whirlwind came out of the north, a glowing cloud, and brilliant fire flashing itself in a circle; and in the midst thereof, an appearance of polished metal (or, gleaming electrum), in the midst of the fire.*

> *As for the likeness of the living creatures, their appearance was like burning coals of fire, and like*

Aliens are Already Among Us

the appearance of lamps: it went up and down among the living creatures. (Ezek 1:13)

This sighting also seems to indicate that he was abducted as he later reports rising into the air.
e. Details of a fresco entitled "The Crucifixion"

Details of a fresco entitled "The Crucifixion". They are located above the altar at the Visoki Decani Monastery in Kosovo. The year it was painted was 1380 A.D.

Diego Cuoghi identified the UFO: it looks just like symbolic and personified representations of the sun and the moon in numerous other paintings of the time.

Aliens are Already Among Us

f. The Dogon Tribe in Africa

In Mali, West Africa, lives a tribe of people called the Dogon. The Dogon are believed to be of Egyptian decent and their astronomical lore goes back thousands of years to 3,200 BC.

According to their traditions, the star Sirius has a companion star which is invisible to the human eye. This companion star has a 50 year elliptical orbit around the visible Sirius and is extremely heavy. It also rotates on its axis.

This legend might be of little interest to anybody but the two French anthropologists, Marcel Griaule and Germain Dieterlen, who recorded it from four Dogon priests in the 1930's. Of little interest except that it is exactly true.

How did a people who lacked any kind of astronomical devices know so much about an invisible star? The star, which scientists call Sirius B, wasn't even photographed until it was done by a large telescope in 1970.

The Dogon stories explain that also. According to their oral

Aliens are Already Among Us

traditions, a race people from the Sirius system called the Nommos visited Earth thousands of years ago. They also appear in Babylonian, Accadian, and Sumerian myths. The Egyptian Goddess Isis, who is sometimes depicted as a mermaid, is also linked with the star Sirius.

The Nommos, according to the Dogon legend, lived on a planet that orbits another star in the Sirius system. They landed on Earth in an "ark" that made a spinning decent to the ground with great noise and wind. It was the Nommos that gave the Dogon the knowledge about Sirius B.

It doesn't seem to explain a 400-year old Dogon artifact that apparently depicts the Sirius configuration nor the ceremonies held by the Dogon since the 13th century to celebrate the cycle of Sirius A and B. It also doesn't explain how the Dogons knew about the super-density of Sirius B, a fact only discovered a few years before the anthropologists recorded the Dogon stories.

The Dogons are a people well known by their cosmogony, their esotericism, their myths and legends that interest foreigners at the highest point in search for culture or tourism. The population is assessed to be about 300,000 people living in the South West of the Niger loop in the region of Mopti in Mali (Bandiagara, Koro, Banka), near Douentza and part of the North of Burkina (North west of Ouahigouya).

The Dogon's (Mali, Africa) homeland has been designated a World Heritage site for its cultural and natural significance. They are also famous for their artistic abilities and vast knowledge about astrology, especially the Sirius star, which is the center of their religious teachings. The Dogons know that Sirius A, the brightest system in our firmament, is next to a small white dwarf called Sirius B,

Aliens are Already Among Us

which was not identified by western scientists until 1978. The Dogons knew about it at least 1000 years ago. Sirius B has formed the basis of the holiest Dogon beliefs since antiquity.

Western astronomers did not discover the star until the middle of the nineteenth century, and it wasn't even photographed until 1970. The Dogons go as far as describing a third star in the Sirius system, called "Emme Ya" that, to date, has not been identified by astronomers. In addition to their knowledge of Sirius B, the Dogon mythology includes Saturn's rings and Jupiter's four major moons. They have four calendars, for the Sun, Moon, Sirius, and Venus, and have long known that planets orbit the sun.

g. UFO Sightings by the Puritans

During the 1600s, Puritans in New England spotted more than just witches flying through the skies. Hundreds of years before Area 51 and Project Blue Book, Massachusetts Bay Colony founder John Winthrop detailed instances of unidentified flying objects in the heavens above seventeenth-century Boston in the first recorded UFO sightings in America.

On March 1, 1639, John Winthrop opened his diary in which he recorded the trials and triumphs of his fellow Puritans as they made a new life in a new land. As the governor of the Massachusetts Bay Colony put pen to paper, he began to recount a most unusual event that had recently caused a stir among the English immigrants.

Aliens are Already Among Us

Portrait of John Winthrop

Winthrop wrote that earlier in the year James Everell, "a sober, discreet man," and two others had been rowing a boat in the Muddy River, which flowed through swampland and emptied into a tidal basin in the Charles River, when they saw a great light in the nighttime sky.

"When it stood still, it flamed up, and was about three yards square," the governor reported, "when it ran, it was contracted into the figure of a swine." Over the course of two to three hours, the boatmen said that the mysterious light "ran as swift as an arrow" darting back and forth between them and the village of Charlestown, a distance of approximately two miles. "Diverse other credible persons saw the same light, after, about the same place," Winthrop added.

The governor wrote that when the strange apparition finally faded away, the three Puritans in the boat were stunned to find themselves one mile upstream—as if the light had transported them there. The men had no memory of their rowing against the tide, although it's possible they could

Aliens are Already Among Us

have been carried by the wind or a reverse tidal flow. "The mysterious repositioning of the boat could suggest that they were unaware of part of their experience. Some researchers would interpret this as a possible alien abduction if it happened today," write Jacques Vallee and Chris Aubeck in "Wonders in the Sky: Unexplained Aerial Objects from Antiquity to Modern Times."

Some have speculated that the curious glow could have been an "ignis fatuus," a pale light that can appear over marshland at nighttime due to the combustion of gas from decomposed organic matter. If Winthrop's report was correct, however, the light was not rising from the swamp but shooting across the sky, making that explanation unlikely.

An odd sight returned to the skies of Boston five years later, according to another entry in Winthrop's diary dated January 18, 1644. "About midnight, three men, coming in a boat to Boston, saw two lights arise out of the water near the north point of the town cove, in form like a man, and went at a small distance to the town, and so to the south point, and there vanished away."

A week later, Winthrop wrote, another unexplained celestial event occurred over Boston Harbor. "A light like the moon arose about the N.E. point in Boston, and met the former at Nottles Island, and there they closed in one, and then parted, and closed and parted diverse times, and so went over the hill in the island and vanished. Sometimes they shot out flames and sometimes sparkles. This was

Aliens are Already Among Us

about eight of the clock in the evening, and was seen by many.

"About the same time a voice was heard upon the water between Boston and Dorchester, calling out in a most dreadful manner, 'Boy! Boy! Come away! Come away!'; and it suddenly shifted from one place to another a great distance, about twenty times. It was heard by diverse godly persons. About 14 days after, the same voice in the same dreadful manner was heard by others on the other side of the town towards Nottles Island."

Unlike the 1639 UFO, Winthrop had an explanation for the latest luminescence over his "city upon a hill." The governor noted the bizarre spectacle was seen near the location where a vessel captained by John Chaddock exploded months earlier after a sailor accidentally ignited gunpowder aboard the ship. The captain was not aboard at the time, but the blast killed five crew members.

Winthrop noted that rescuers had recovered the bodies of all the victims except for the man believed responsible for the calamity, a sailor who professed the ability to communicate with the dead and was suspected of murdering his master in Virginia. The hand of the devil was thought to have taken possession of the body, and it was the haunting voice of the sailor's ghost that was said to have accompanied the strange vision of Ye Olde UFO that mystified Boston.

Aliens are Already Among Us

h. Sightings in the Nineteenth Century

There are also sightings in more recent history so these mysterious events aren't just limited to ancient reports. Tales of alien spaceships on Earth appeared in 19th Century papers:

> "About 35 miles northwest of Benkelman, Dundy County, on the 6th of June [1884] a very startling phenomenon occurred. It seems that John W. Ellis and three of his herdsmen and a number of other cowboys were out engaged in a round-up. They were startled by a terrific whirring noise over their heads, and turning their eyes saw a blazing body falling like a shot to earth. It struck beyond them, being hidden from view by a bank."

> The article, from the Nebraska Nugget, goes on to say that the rancher found "fragments of cogwheels, and other pieces of machinery" lying on the ground. The heat was so intense that "as to scorch the grass for a long distance around each fragment and make it impossible for one to approach..." The group found the main part of the wreck and one of them "fell senseless from the gazing at it at too close quarters. His face was blistered, and his hair singed to a crisp."

> "Finding it impossible to approach the mysterious visitor [the UFO] the party turned back on its trail. When it [the UFO] first touched the earth the ground was sandy and bare of grass. The sand was fused to an unknown depth over a space about 20 feet wide by 30 feet long, and the melted stuff was still bubbling and hissing."

Aliens are Already Among Us

Later in the story the ship is described as being 50 to 60 feet long, cylindrical and 10 to 12 feet in diameter. The writer notes that it was apparently composed of metal with an appearance like brass, but was remarkably light. The story also notes that the wreck is located in a remote and wild region and "the roads are hardly more than trails."

The second story appeared in the *Dallas Morning News* on April 19th, 1897:

> "About 6 o'clock this morning the early risers of Aurora were astonished at the sudden appearance of the airship that had been sailing throughout the country. It was traveling due north, and much nearer the earth than before."
>
> The article describes how the air vehicle "sailed over the public square and when it reached the north part of town collided with the tower of Judge Proctor's windmill and went to pieces with a terrific explosion, scattering debris over several acres of ground, wrecking the windmill and water tank and destroying the judges' flower garden." It continues with, "The pilot of the ship is supposed to have been the only one aboard, and while his remains are badly disfigured, enough of the original has been picked up to show he was not an inhabitant of this world."
>
> "Mr. T. J Weems, the U.S. Signal Service officer at this place and an authority on astronomy, gives it as his opinion that he [the pilot] was a native of Mars."
>
> According to the story the remains of the ship were composed of a strange metal that seemed a

Aliens are Already Among Us

mixture of aluminum and silver. The townspeople came to view the wreck and pick up specimens. The pilot was buried the day after the article was published.

There are many more examples of mysterious and anomalous evidence that some type of beings visited the earth in our past. There are enough examples here since there are many more books on historical UFO and Alien records.

My purpose here was just to show that this subject didn't just start recently but has been a controversial subject for all of human history.

4.0 Roswell and EBEs

Ahh… The Roswell Incident. How many thousands of articles and books have been written about the likely alien ship crash near Roswell, New Mexico in 1947. There is now even a museum there to visit there.

Here is one famous newspaper issued after the crash before the military changed their story to say a weather balloon had crashed:

Most retellings of this story include that dead alien bodies were recovered. Some stories are that the aliens were alive and were kept in isolated conditions for years.

How to separate fact from fiction? I think it's fair to say that much of the UFO culture around the world is based on this incident. I have read several books which talked about testimonies of aliens captured from the accident, technology developed as a result, and more.

All I can say for sure is that this is the biggest event in recorded UFO history so it's likely there was some basis in fact.

Aliens are Already Among Us

Here is a story about an EBE supposedly found in the Roswell crash as posted on serpo.org:

Section I: EBE #1: The UNTOLD STORY OF ITS LIFE AFTER THE ROSWELL CRASH OF JULY 1947 ...

-- As Relayed by Anonymous II --

Ebe #I — Was found in the Roswell crash site near Corona, NM. Ebe #I was slightly hurt and quickly recovered from his wounds. Ebe #1 was a mechanic. He was able to communicate through pictures with Army personnel. They placed him in an isolation room at the Roswell Army Airfield. In September 1947, he was transferred to Kirtland Field and was again isolated in a medical unit.

During this time period, Ebe #1 worked with top military linguists. They were able to communicate with him by showing him photographs. They later developed symbols to communicate words.

[In linguistics, these are known as LOGOGRAMS or LOGOGRAPHS].

Eventually, Ebe #I was able to utilize these symbols to communicate his wants and needs. Ebe # I was able to eat simple foods, such as bread, fruits, pasta, salads, and cheese. However, meats caused him problems and he would vomit the meats. Ebe #1 was examined by numerous medical doctors and scientists. They took skin samples, examined his body fluids and conducted X-rays on him. They found that Ebe #I had one (1) primary organ that worked as his heart and lungs; that organ was combined.

Aliens are Already Among Us

He had a simple digestion system. One (1) organ worked as a stomach and one (1) other organ worked as his intestines. They could find NO liver, pancreas, or gall bladder. Apparently, Ebe #l's stomach acted as all of these organs or he just didn't need them. Ebe #l had small glands on his hands, arms and legs. His glands would enlarge at certain times. Scientists could not figure out what these glands were for nor their use.

Ebe #l measured 4'3" tall and weighed 60-lbs. His weight never varied; however his height did. During the winter months, his height would increase by about one (1) inch. His body generated heat by some means that our scientists could not determine. Ebe #l wore a tight-fitting 1-piece suit. That suit was all he needed to maintain his correct body heat. The suit was made of an elastic material that retained warmth and kept out cold. Ebe #l had a body heat of 101 degrees which almost never varied. He was provided with a blanket, but he rarely used it.

Ebe # I's blood was light red, but contained similar cells such as red and white blood cells. This blood also contained numerous things our scientists could NOT identify. Ebe #l's body did not require a large quantity of water/fluids. He was able to extract the fluids required to maintain his proper levels through the break

down of food. His body was able to somehow determine the correct amount of fluids required and eliminate the remaining UNused fluids. It was determined that eating was NOT a pleasurable activity for Ebe #l, but rather a necessity.

Aliens are Already Among Us

Ebe #I was always calm, kind and very considerate. He <u>never</u> got excited, rude nor was mean spirited. He was very social, even though he couldn't understand others around him. He liked to touch people and learned that holding a human hand was a common social practice. He was very docile. Even when scientists were poking and examining him, he seemed to understand as he just let them go about their business. He was always willing to communicate or attempted to communicate. He quickly learned the symbols system and eventually learned our language in a very simple way.

In 1950, Ebe #I was moved to a special facility created for him at the Los Alamos National Laboratory. He lived in a little 3-room apartment. His handler, who was assigned to him in 1949, joined and lived with him in the apartment. They became such great companions that neither wanted to separate.

It was found that Ebe #1 could NOT speak our language because of the lack of vocal cords in his throat [the Ebens communicated via tonal inflections amongst themselves]. A brilliant doctor developed a device that was implanted into Ebe #I's throat which allowed him to speak English. Although crude, he was able to speak simple sentences and eventually communicate through English to his handler.

During the entire time Ebe #1 was alive, he had medical problems. He developed a rash on his body that irritated Ebe #I. Several different medicines were tried and eventually the rash went away. Ebe #1 developed a cough that seemed to

Aliens are Already Among Us

be connected to a food allergy. They found that certain fruits would cause the cough. Eventually, the cough subsided, but he was left with a sore throat.

Ebe #1 was shown the items recovered in the Roswell crash site. Ebe #1 was able to teach us how to utilize the communication and energy devices. We also found a medical kit, which contained small injectable tubes. Ebe #I didn't know what each item was for, but he explained they were for injuries. Scientists experimented and determined that each tube contained a chemical substance. Not knowing what they were for, they were cautious not to use them on Ebe #I for fear they would do more harm than good.

-- As Relayed by Anonymous I [below] --

There was NOT a rescue mission launched for Ebe #1. The only two (2) Eben crafts that were near Earth crashed. Ebe #1 did say they sent a distress signal to Planet SERPO, but it would take at least nine (9) months for the nearest rescue craft to reach them. Ebe #1 died in 1952, and his body, along with those of his crew mates, were returned to the Ebens in 1964 during the meeting in New Mexico. The Eben craft was stored in Ohio and then moved to NV; I believe it is still at the Nevada Test Site, NV.

There were two (2) other exchange missions. I was never involved and don't have the details. The only surviving Eben was taken to the Roswell base and placed in the custody of the lead intelligence officer. The Eben was totally isolated. He was moved to

Aliens are Already Among Us

Los Alamos the next day by a military convoy. The Eben died at Los Alamos in 1952. He was isolated from almost everyone. President Truman saw him. No Navy JAG was ever involved. That is another one of those totally false stories. These UFO "researchers" who write these books and make up stories ... only a few know the actual truth about the Roswell crash.

The Ebens did not have this [teletransporter] technology. They have space travel and can venture through space defying the time barrier. As for our technology: I don't think we have it. We don't have the Star Trek "beam-me-up-Scotty" type of technology.

Aliens are Already Among Us

5.0 Serpo—Missions to Other Planets

One of the most interesting stories in the whole UFO and Alien realm is the story of the Serpo Project. The Serpo project is about how twelve military personnel were selected for a mission to the planet of the Ebens and how they stayed for thirteen years before returning.

For those of you who have seen the movie "Close Encounters of the Third Kind" it's also really interesting how the ideas of a team of military personnel traveling on an alien ship to another star system is a very parallel concept.

Here is the original message posted on serpo.org which gives basic information about the program:

Aliens are Already Among Us

First let me introduce myself. My name is Request Anonymous. I am a retired employee of the U.S. Government. I won't go into any great details about my past, but I was involved in a special program.

As for Roswell, it occurred, but not like the story books tell. There were two crash sites. One southwest of Corona, New Mexico and the second site at Pelona Peak, south of Datil, New Mexico.

The crash involved two extraterrestrial aircraft. The Corona site was found a day later by an archaeology team. This team reported the crash site to the Lincoln County Sheriff's department. A deputy arrived the next day and summoned a state police officer. One live entity [EBE] was found hiding behind a rock. The entity was given water but declined food. The entity was later transferred to Los Alamos.

The information eventually went to Roswell Army Air Field. The site was examined and all evidence was removed. The bodies were taken to Los Alamos National Laboratory because they had a freezing system that allowed the bodies to remain frozen for research. The craft was taken to Roswell and then onto Wright Field, Ohio.

The second site was not discovered until August 1949 by two ranchers. They reported their findings several days later to the sheriff of Catron County, New Mexico. Because of the remote location, it took the sheriff several days to make his way to the crash site. Once at the site, the sheriff took photographs and then drove back to Datil.

Aliens are Already Among Us

Sandia Army Base, Albuquerque, New Mexico was then notified. A recovery team from Sandia took custody of all evidence, including six bodies. The bodies were taken to Sandia Base, but later transferred to Los Alamos.

The live entity established communications with us and provided us with a location of his home planet. The entity remained alive until 1952, when he died. But before his death, he provided us with a full explanation of the items found inside the two crafts. One item was a communication device. The entity was allowed to make contact with his planet.

Somehow, I never knew this information, but a meeting date was set for April 1964 near Alamogordo New Mexico. The Aliens landed and retrieved the bodies of their dead comrades. Information was exchanged. Communication was in English. The aliens had a translation device.

In 1965, we had an exchange program with the aliens. We carefully selected 12 military personnel; ten men and two women. They were trained, vetted and carefully removed from the military system. The 12 were skilled in various specialties.

Near the northern part of the Nevada Test Site, the aliens landed and the 12 Americans left. One entity was left on Earth. The original plan was for our 12 people to stay 10 years and then return to Earth.

But something went wrong. The 12 remained until 1978, when they were returned to the same location in Nevada. Seven men and one woman returned. Two died on the alien's home planet. Four

Aliens are Already Among Us

others decided to remain, according to the returnees. Of the eight that returned, all have died. The last survivor died in 2002.

[Clarification (BR/ Victor Martinez): the paragraph above contains a typo in the original. Twelve team members went, and eight returned – two having died on Serpo and two having chosen to remain; these two were not ordered to return.]

The returnees were isolated from 1978 until 1984 at various military installations. The Air Force Office of Special Investigation (AFOSI) was responsible for their security and safety. AFOSI also conducted debriefing sessions with the returnees.

I have never seen or read anything about the exchange program. I once heard a little bit of information from Linda Howe, but she didn't have much information.

I've monitored your e-mails for about six months. I've read e-mails from you and others. But I've never seen nor heard the truth about the real Roswell incident or the exchange program.

I'd like to hear what others say about this.

(See serpo.org for all of the rest of the details)

*****

Below is a picture of one of the alien spaceships:

Aliens are Already Among Us

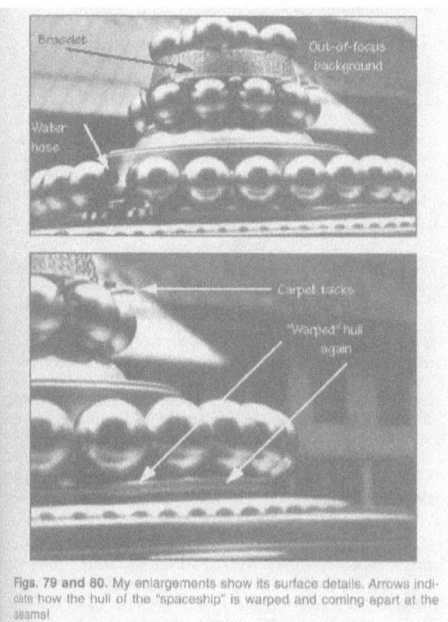

Figs. 79 and 80. My enlargements show its surface details. Arrows indicate how the hull of the "spaceship" is warped and coming apart at the seams!

Another tie in is the internal consistency of the reports among themselves and the tie in to other Alien and UFO information. For instance, one the first alien abduction cases created a star map under hypnosis which showed the star Zeta Reticuli (Betty & Barney Hill- as discussed in the abduction chapter of their book) as where the aliens were from. This is the same star system that humans are supposed to have visited per Serpo.org.

All of the information sourced here is from one website: www.serpo.org

Aliens are Already Among Us

Aliens are Already Among Us

6.0 Alien Bases on Earth

The story of alien bases on Earth, or under the Earth, or even under the ocean is a whole area of discussion by itself.

a. <u>Indian Springs, California</u>

The book "Millennial Hospitality" By Charles James Hall describes the experiences of an Air Force Weatherman who was stationed in the California Desert.

This book describes (in several volumes) how Charles met some Aliens he calls "The Tall Whites" who had a hidden base in the desert and came to Earth to use it as a transit point to other stars.

(Drawing of Tall Whites)

Aliens are Already Among Us

They even visited Las Vegas occasionally and liked to go to the "Stardust Casino" to see the Space decorations. Charles describes his interactions with them in detail over the course of several books.

He also had a reader send a really interesting email to him as follows:

> *Thomas Finley from Essex, England writes, "Before I even read any of your books and what you had encountered in the desert in the 1960s, I shared some letters and e-mails with an ex-state department official that was assigned to a group of these same beings in the Washington DC area in the early 1950s. When I first started to share letters with this gentleman, I was only looking for others that shared interest in UFOs, and those that were in the military.*

Charles further comments on the emails:

> *In his emails, he told me his small cadre Of secret service and a mixture Of military affairs personnel had the task of providing care to a group Of tall white aliens that were guests of the U.S. Government in the D.C. area. One of the beings he told me a great deal of was a young female that was called, the "Princess" He related that she would like to go for long night drives around the city, and on occasion, go out in disguise and meet regular humans." One of the most interesting things about Tom Finley's letter is that his friend was taking care of a group of tall white aliens in the early 50s. I was very pleased,*

Aliens are Already Among Us

as I had earlier received a copy of a January 5, 1951 Las Vegas Review Journal newspaper clipping from Bob Wood.

The article said that the government was spending $300,000,000 on housing facilities! It makes perfect sense that such a sum in 1951 dollars was indeed needed to build the underground living quarters and hangers for the Tall Whites at Indian Springs, and that the Tall Whites were guests in the Washington D.C. area during the construction.

Aliens are Already Among Us

b. The Dulce Base

The greatest claim about an alien base is the one for the underground base at Dulce, New Mexico.

Starting in 1979, Bennewitz became convinced he was intercepting electronic communications from alien spacecraft and installations outside of Albuquerque. By the 1980s he believed he had discovered an underground base near Dulce. The story spread rapidly within the UFO community and by 1990, UFOlogist John Lear claimed he had independent confirmations of the base's existence.

Political scientist Michael Barkun writes that Cold War underground missile installations in the area gave superficial plausibility to the rumors, making the Dulce base story an "attractive legend" within Ufology. According to Barkun, claims about experiments on abductees and firefights between aliens and the Delta Force place the Dulce legend "well outside even the most far-fetched reports of secret underground bases."

Some claim that there is even an underground tunnel network connecting bases internationally:

> *An underground Military Base/Laboratory in Dulce, New Mexico connects with the underground network of tunnels which honeycombs our planet, and the lower levels of this base are allegedly under the control of Inner Earth beings or Aliens. This base is connected to Los Alamos research facilities via an underground "tube-shuttle." (It can be assumed that such a shuttle way would be a straight-line construction.*

Aliens are Already Among Us

It should then be possible, by using maps and some deduction, to determine the most likely location of this base, especially since the general location is already known.) Beginning in 1947, a road was built near the Dulce Base, under the cover of a lumber company. No lumber was ever hauled, and the road was later destroyed. Navajo Dam is the Dulce Base's main source of power, though a second source is in El Vado (which is also another entrance). (Note: The above facts should also help to locate the base.) Most of the lakes near Dulce were made via government grants "for" the Indians.

(Note: The September, 1983 issue of Omni (Pg. 80) has a color drawing of 'The Subterrene,' the Los Alamos nuclear-powered tunnel machine that burrows through the rock, deep underground, by heating whatever stone it encounters into molten rock, which cools after the Subterrene has moved on. The result is a tunnel with a smooth, glazing lining.)" (Note: Where would the molten rock go? And what has been done with this concept since 1983?)

Below is a diagram of possible underground tunnels connecting various secret bases together:

Aliens are Already Among Us

Aliens are Already Among Us

c. Additional Alien Underground Bases:

There have been many claims of alleged underground bases in different parts of the planet, most of which are believed to be serving to the aliens as their experimental or military power testing grounds. Here is a list of such top 10 underground alien bases which are feared as the hub of alien operations on Earth.

Underground alien bases in Europe

(1) Initially commissioned as the official development facility of gas mask technology during World War I, the Porton Down base in Wiltshire, England is believed to be one of the most active underground alien bases of the world. The fact that Wiltshire gets maximum share of crop circles being made on its fields supports that fact that aliens are very active in this part of the world.

(Porton Down base in Wiltshire, England)

Aliens are Already Among Us

(2) A large iron plate was discovered by a laborer in the fields of Staffordshire in England, the hatch of whose was oval and large with an iron ring mounted on it. The plate served as a closed gate to a tunnel which is believed to connect to an underground alien base. The whole thing is mentioned in the 18th Century historical documentary book titled 'A history of Staffordshire'.

(Staffordshire Tunnel, Underground Base)

Underground alien bases in Canada

(3) One of the iron mines in Newfoundland was being shut down owing to various strange events that occurred in the mine when it was dug deeper than the other mines in that area. As per eye witnesses, shadows of beings of unusual shape and structure were being spotted and various mysterious things happened inside the mine, which compelled people to believe that it was serving as an underground alien base on earth.

Aliens are Already Among Us

Underground Base of the iron mines in Newfoundland

(4) The Valley of the headless man or the Nahanni Valley in Canada is a mysterious piece of land as people entering the valley are mostly founded dead with their heads being severed brutally. Multiple such incidences have infused the fear in the minds of the locals that this area is being captured by a group of hostile aliens who cause such vicious death to anyone who enters 'their' area.

Aliens are Already Among Us

(Mysterious Nahanni Valley in Canada)

Underground alien bases in Russia

(5) Russia's Roswell or the Zhitkur Base in Russia is a hub of different alien operations that are being carried out by extraterrestrial beings. Officially, the base is supposed to serve as a rocket launch facility, but the reality is that all sorts of secret researches by aliens are conducted under the snows.

Aliens are Already Among Us

(Underground base of Zhitkur Base in Russia taken from Google earth)

<u>Underground alien bases in America</u>

(6) If Hopi legends are to be believed, the caves near the confluence of Little Colorado and Colorado rivers used to be underground homes of ant people who were friendly with local humans. Some of these caves have been explored to discover Central American, Egyptian and Oriental type artifacts which has made people believe that this area served as a secret underground alien base.

Aliens are Already Among Us

Caves in Colorado

(7) As per local folklore there are many tunnels beneath Mt. Shasta in California which leads to an underground alien base. The fact that California is one of the states that has maximum UFO sightings makes it easy to believe that the UFOs come to meet up with the aliens stationed at the underground base beneath Mt. Shasta.

Aliens are Already Among Us

Mt. Shasta underground base

(8) Another entrance to a large underground city can be found near Mt. Lassen in California's Tehama County. The residents of Manten, the city at the foot of Mt. Lassen have known about this underground base for long and believe it to be an operational hub of alien visitors.

Aliens are Already Among Us

Cave entrance of Mt. Lassen

(9) The most famous underground alien base is the Dulce Base in New Mexico. This underground base is believed to have multiple lairs, with each lair being the ground of horrific genetic experiments on humans and other forms of life on the earth. A population of 18,000 aliens is believed to inhabit the Dulce Base.

Aliens are Already Among Us

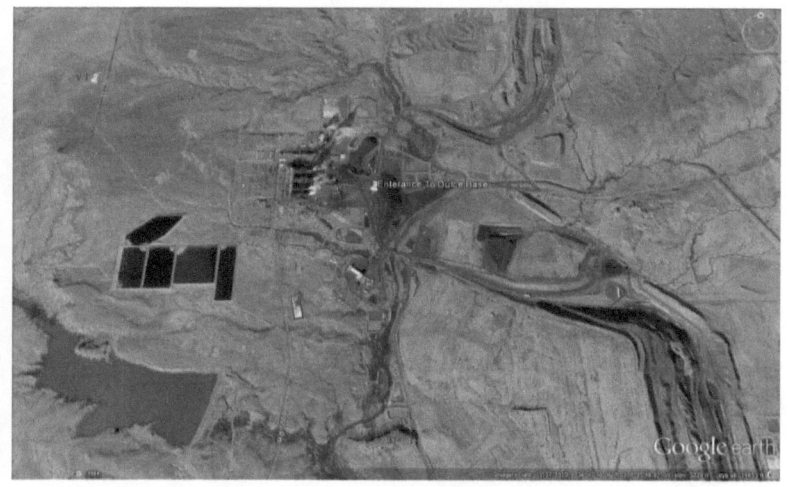

Dulce Base in New Mexico taken from Google earth

(What is going on at Dulce is described above)

(10) What is being covered up as a secret military air force research and testing grounds is actually an underground alien base where captivated aliens from the 1947 Roswell UFO crash are kept and studied. We are talking of Area 51 located in the deserts of Southern Nevada, the most discussed and most hyped alien base under the soils. It is believed that alien spacecrafts and alien corpses are being studied here and reverse engineering is also being conducted on the alien spaceships.

Aliens are Already Among Us

Area 51 underground alien base

Aliens are Already Among Us

7.0 Aliens Underwater

It's not just on land that Alien Spacecraft have been observed. Weird sightings in the Oceans of the world have also been made with alien craft entering and exiting the water. Sometimes referred to as USOs (Unidentified Submerged Objects) or UUOs (Unidentified Underwater Objects), sightings of strange lights and crafts emerging from, or descending into, the seas, oceans, and lakes around the world are quite widespread.

Some UFO researchers and investigators even claim that there are underwater alien bases present in the vicinity of these aquatic sightings. Here are 10 of the most interesting claims.

a. <u>Underwater Base off The Coast Of Malibu</u>

California has been a UFO hot spot for years, so when it was theorized that a structure discovered 600 meters (2,000 ft) below the water 10 kilometers (6 mi) off the coast of Malibu was in fact an alien base, it didn't come as that much of a surprise. The structure has a distinct oval shape

Aliens are Already Among Us

and what appears to be flat top. It measures approximately 5 kilometers (3 mi) wide, and there also seem to be pillars holding up the main "roof" of the mysterious anomaly. These pillars look to be spaced evenly apart, almost as if it were an entrance of sorts.

The alleged alien base was first brought into the public arena when Jimmy Church, who hosts the radio show Fade To Black, acted upon information given to him by a listener, who went by the name of "Maxwell." Maxwell gave coordinates from Google Earth that showed something he believed may be an artificial structure. Church, with his design team, obtained as many different angles of the structure as they could. Church stated that after investigating all of these angles, he very much believes this structure is indeed artificial, even going as far as to say it could be the "holy grail" for ufologists.

Aliens are Already Among Us

b. <u>Underwater Base Off Puffin Island Wales</u>

Numerous sightings of strange lights leaving and entering the sea around Puffin Island, Wales, has led some ufologists to believe that the area may be home to an underwater alien base. A spate of sightings in early 1974 were particularly interesting to investigators. All spoke of not just lights, but solid objects that were seen leaving the sea near the island.

UFO investigator Phil Hoyle stated that he has heard and read various unconnected reports concerning Puffin Island and that all of them tell the same story and describe the same type of phenomenon. He also stated that alleged alien abductees close to the area have reported that their abductors were humanoid and told them they came from a base under the sea near the island.

There are even some theories that the area itself may be the ancient legendary kingdom of Cantre'r Gwaelod and that these humanoids are their descendants.

Aliens are Already Among Us

Hoyle believes that the ties to an ancient site and the current UFO activity shouldn't be downplayed, stating that according to his research, there is an 80 percent increase in such activity on or around ancient sites worldwide.

c. UFO Base Under Lake Ontario

In December 2013, MUFON published a report from a Hamilton resident who claimed to have seen several strange, glowing orbs hanging over Lake Ontario. He also claimed to have seen these orbs on the lake several times prior, and what's more, he believes that there is an alien base under the water. He is far from the first person to have made such an assertion.

In August 1981, five witnesses who were driving alongside Lake Ontario early one evening saw a dome-shaped craft flying over the water. They followed the craft for some time before seeing it begin to descend and enter the water, disappearing from their sight. Perhaps the stories of an underwater alien base, stem from the 1977 book The Great Lakes Triangle by Jay Gourley, who made note that many planes and people had disappeared over Lake Ontario, not to mention the many UFO sightings of the area.

Another book, Underground Alien Bases, released in 2012 and written by the somewhat strange "Commander X," has also perpetuated the legend of the alien base under the lake. It features within its pages several accounts of sightings on and around the lake and the assertion that an alien fortress of one kind or another lies under the water. However, none of the accounts can be verified by a secondary source and so are left open to debate as to how reliable they are.

Aliens are Already Among Us

d. <u>Flight Corridor to Underwater Base North Island, New Zealand</u>

North Island in New Zealand has been witness to many strange objects entering the sea, with activity stretching out to some of the offshore islands around the area. UFO researchers state that the area is a hot spot for such activity, with locals claiming that the area is a flight corridor to an underwater base for the strange craft that exist there. Between January and March 1995, there were dozens of sightings of these objects coming and going in the area. These were not limited to nighttime, with many of the objects being witnessed during the daylight hours.

One particular daylight sighting on March 9, 1995, stands out. It began with two fishermen who witnessed a bright, silver, ball-shaped craft that seemed to glow or pulse as it moved and emitted a strange red stream behind it. The two men saw the object for less than 10 seconds before it vanished from their sight. Around two minutes later, however, there were sightings of what seemed to be the same object by the control towers at both Hamilton International and Rotorua Airports. Further supporting the incident were several reports that were phoned in to a local radio station from concerned residents, who all described a very similar object in the bright, sunny sky.

   e. Pacific Ocean 'Humming' Is Actually Marine Life
      Breaking Wind

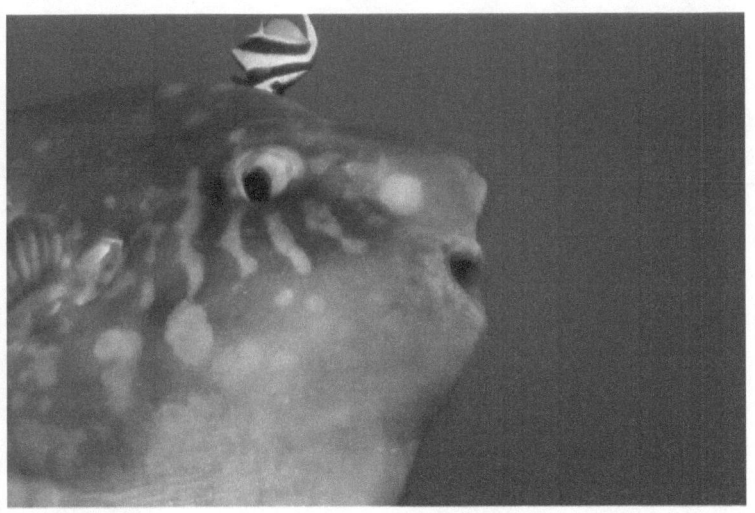

In the case of the alleged underwater alien base in the depths of the Pacific Ocean, the mainstream scientific explanation is actually stranger than that of the ufologists. According to a research team led by Simon Baumann-Pickering, the low humming sound that has had people searching for bases and installations for over two decades is nothing more than sea creatures releasing gases from their swim bladders—essentially breaking wind.

UFO researchers have almost entirely rejected this assertion.

Aliens are Already Among Us

The humming has been studied and debated since 1991 by scientists and UFO investigators alike, when it was first observed by National Oceanic and Atmospheric Administration (NOAA). Ufologists argue that rather than being a sound of the natural world, the sounds are more akin to those of electrically powered artificial structures. To them, this is proof of the existence of top secret underwater bases, which are quite probably extraterrestrial, given the advanced technology that would be required to build and operate such an underwater building.

Other theories include that the humming could be generated by marine vessels or an unknown geological phenomena.

f. <u>Strange Lights and Sounds in Puerto Rico</u>

The island of Puerto Rico has had numerous accounts of UFO sightings, particularly along its northeastern coastline, where strange craft have been seen both emerging out of

Aliens are Already Among Us

and disappearing into the water for years. There is also a strong US Navy presence on the island, and stories from locals and US expats alike state that the US military at the very least is monitoring the strange activity on the waters along the coast.

Experienced ufologist and researcher Nick Redfern recently told of an ex–civil defense employee who had witnessed a strange craft emerge from the waters off the Puerto Rican coast. It steadily rose up into the air before shooting off at speed. Another account came to Redfern from a local police officer, who stated that the US Navy had spent considerable time tracking an underwater craft along the northeastern coast of the island.

There have been numerous claims and theories that the immediate coastline of this Caribbean island is home to underwater alien bases, perhaps none more specific than those of investigative journalist Jorge Martin. Martin stated that using NOAA satellite images, they found several anomalies around the island which suggested artificial structures. Martin stated that these structures were unusually large and precisely rectangular and were generally on the eastern and southern coast of the island. He also claimed that coming off many of these structures, there appeared to be what looked like tunnels leading to other structures and even to the mainland.

What is interesting is that in the Puerto Rican city of Ponce during the late 1980s, local residents were complaining to authorities that they could hear loud rumbling, which seemed to be coming from beneath the ground. It was reported that the sounds were similar to those you might expect to hear when heavy machinery was operated. The sounds seemed to stop after several days. One of the places on the mainland that Martin's tunnels appeared to connect to was the city of Ponce.

g. <u>Ancient Alien Base at Lake Titicaca</u>

The predominantly still waters of Lake Titicaca on the borders of Peru and Bolivia are not only the world's highest navigable waters, but also host to a plethora of UFO activity and according to some, home to an advanced and possibly ancient alien base.

The ancient city of Tiwanaku, considered to be one of the oldest cities in the world, sits on the southeastern shore of Lake Titicaca. Ufologists, ancient astronaut theorists in particular, state that the advanced level of agriculture, irrigation, and astronomy as well structures that suggest advanced building techniques leads them to believe that an extraterrestrial race once resided here. Numerous texts and statues have also been found in the immediate vicinity of the lake, which seem to depict ancient Mesopotamian underwater gods—gods that they claim were from an alien civilization that once resided under the still waters of Lake Titicaca.

Aliens are Already Among Us

Recently, a video surfaced on YouTube shot by a group of Italians who were on a spiritual journey of sorts in the region. As they pointed their camera out to the still, blue waters of Lake Titicaca, what appeared to be a large submerged object was seen slowly making its way away from the shore. It appeared to be disc-shaped object, and as it moved through the water, its shape remained the same, suggesting that it is indeed solid.

If there was an ancient alien underwater base below the surface of Lake Titicaca, perhaps this video may suggest that it is very much still in use.

h. <u>Underwater Alien Base at Guantanamo Bay</u>

According to a former US Marine who served at Guantanamo Bay in the late 1960s, there is an underwater alien base off the Cuban coast. Furthermore, the Marine claimed that of the many UFO sightings of strange objects going in and out of the water, the US military has even managed to capture several on film. He also states that he and his colleagues were under strict instructions not to talk about the strange activity they witnessed there.

The Marine claimed that the craft he witnessed appeared to be made of a "dull" metal with a series of blue lights. When they would enter the water, the blue lights could still be witnessed but would grow fainter and fainter, suggesting to him that whatever the object was, it was descending deeper into the water.

Perhaps coincidentally, around 140 kilometers (90 mi) north of the US base in Gulf Breeze, Florida, there have been several UFO sightings that seem to match the descriptions of the craft as described by the unnamed marine, and may suggest they are originating from the same place.

One particular sighting came in November 1987, when Ed Walters claimed that he witnessed a strange object fly overhead from the coast. It emitted a bright, blue light that "trapped" him in its beam. He stated that while he was caught in it, everything around him was blue, and he could not move. Walters did manage to take several pictures of the craft. Hundreds of witnesses came forward to say they, too, had seen the strange object that evening.

There is debate over the credibility of his account. MUFON seemed to believe the sighting was genuine, while others have stated that the photographs were obvious fakes. In a strange twist, years after Walters' claims, an investigation into his story was said to have discovered a model craft very similar to the object that Walters stated he had photographed in a house where he once resided.

Supporters of his claims, however, argue that this model is an obvious "plant" to discredit him.

i. <u>Increased UFO Sightings At Lake Erie Cleveland, Ohio</u>

Lake Erie, one of the five great lakes of North America, has a long history of strange lights and unexplained phenomena. However, the area around Cleveland, Ohio, has experienced an increase in UFO sightings over the years.

The 2011 the book Eerie Erie: Tales of the Unexplained from Northwest Pennsylvania by Robin Swope examines and investigates numerous MUFON reports. Many of these reports speak of UFOs seen "crashing" into the water of Lake Erie. One particular reports from 1988 alleged a strange craft landed on the lake when it was iced over. The landing was witnessed by Sheila and Henry Baker, who made a report to the local Coast Guard. As the mysterious craft landed, there were strange sounds heard coming from the ice as well as a series of blue and red lights from

Aliens are Already Among Us

the craft itself. There also appeared to be several strange triangular objects jettisoned from the descending object. These triangles seemed to move purposely around in all directions along the icy surface of the lake. Suddenly, the sounds on the ice stopped, and the craft and the mysterious triangles vanished, suggesting that they had indeed found their way below the ice and under the water.

According to a report in 2007 that ran on News Channel 5 in Cleveland, the area had seen no less than 20 UFO sightings in only two years. There are a great many videos on YouTube that claim to show some of these UFOs over Cleveland. Local UFO researcher Richard Lee stated that the UFOs seem to take particular interest in any new building projects undertaken in the area. Once a new project begins, according to Lee, there is usually a UFO sighting in the immediate vicinity not long after. Are they coming from their alleged base in Lake Erie?

Aliens are Already Among Us

j. <u>Lake Baikal—Scene of an Underwater Battle with Aliens</u>

The ancient Lake Baikal in Siberia is said to be the deepest lake on the planet and has a history of UFO activity as well as alleged sightings of aliens underneath the water by Russian military divers.

These sightings date back to the Soviet era, but since the end of the Cold War, the reports have come into the public domain. Many of the accounts tell of a huge "mothership" hovering above the expanse of water and even of humanoid beings in strange, shiny suits climbing down from these ships and into the water.

Former Soviet naval officer and ufologist Vladimir Azhazha claimed to have leaked top secret files relating to an incident in 1982. He stated that military divers, who were on standard training exercises in the area, witnessed huge underwater craft that moved with a speed they had never seen before.

Several days later, the story took an even more bizarre twist when divers witnessed several strange beings under the water. They wore shiny suits and what appeared to be a small, advanced oxygen mask. According to the alleged leaked documents, the unit was ordered to capture these strange creatures. When they attempted to do so, however, they were fought off with what appeared to be an advanced sonar wave weapon that ultimately killed three of the seven divers. The remaining four, now terrified and injured from their attempts detain the strange crew, retreated and made their report to superiors.

In 2009, photographs taken from aboard the International Space Station seemed to show two distinct saucer-shaped anomalies in the region, fueling further speculation that an extraterrestrial presence was under Lake Baikal and even the possibility that these "saucers" were the craft that had been witnessed nearly three decades earlier by the Soviet diving unit.

Aliens are Already Among Us

8.0 NATO's Top Secret Study

The story of the Top Secret study done by NATO about Aliens in 1964 is a fascinating story. The full story is presented here as I found it (in italics):

NATO Meets E.T.

Name: Robert O. Dean, retired Army command sergeant major

Claim: Back in the Sixties, NATO issued a classified report stating that UFOs were real, of extraterrestrial origin, and had visited the earth. This extraordinary report was said to come out of NATO's command center, the Supreme Headquarters Allied Powers, Europe (SHAPE), located then just outside of Paris, France.

Background: Dean, a highly decorated veteran, served on the front lines in both Korea and Vietnam. In 1963, while assigned to the Supreme Headquarters Operations Center (SHOC), SHAPE's war room, headed up by then-supreme allied commander of Europe, Gen. Lyman Lemnitzer, Dean claims he was able to read the detailed 12-inch-thick NATO report on UFOs.

The Story: "SHAPE was one of those choice assignments. You had to have a spotless record and pass security background checks. I applied on a whim and got it. I was very proud and pleased. At SHAPE, I was put through more security checks, given a Cosmic Top Secret (yes, this is a real term) clearance, the highest NATO has, and assigned to

Aliens are Already Among Us

the Supreme Headquarters Operations Center, known as SHOC, the NATO war room. In those days, the activity would run hot and cold and much of it would depend on how the Soviets wanted to play it. The most intriguing thing to me was that we were continually having a problem with large, metallic, circular objects that would appear over central Europe; these were reported as visual phenomena by our pilots and appeared on radar as well. Some flew in formation, and most of the time we spotted them coming out of the Soviet Union, over East Germany, West Germany, France, and then they would often circle somewhere over the English Channel and head north, disappearing from NATO radar over the Norwegian Sea. These objects were very large, moving very fast, at very high altitudes--higher than we could reach at the time--and they seemed obviously under intelligent control

"I was told this had been going on for some time and that in February 1961 there had been quite a scare. Fifty of these objects were spotted on radar and headed in formation from the Soviet Union toward Europe, flying at about 100,000 feet. The Soviets had closed all borders. Everybody went to red alert. All hell broke loose. We really thought `The War' had started. We scrambled. We knew the Russians were scrambling. It was the largest number of these objects that had been seen. Fortunately--and only by the grace of God--we didn't start bombing and neither did the Russians. In nine minutes, they were gone.

"I was told that then-Deputy Supreme Allied Commander of Europe, Sir Thomas Pike, had been

Aliens are Already Among Us

repeatedly requesting information from London and Washington about these objects, but nothing would ever come. We found out later that the Columbine-Topaz spy ring in Paris was intercepting everything and forwarding it to the KGB, which often got intelligence information even before we did. So Pike decided, I was told, to develop an in-house study to determine whether these objects were a military threat.

"In the meantime, the UFO matter literally brought about the establishment of direct communication between the East and West in 1962, which I have always found interesting and ironic. We had pretty well determined by that time that these were not Russian craft, and the Russians had determined they were not ours. So, we came to an understanding, and a direct telephone line was opened between SHOC and the Warsaw Pact Headquarters Command. Of course, a setup was always a possibility, so we had backup ways of checking out whether the Russians were being truthful. But since we were both armed to the teeth and World War III was just ticking away, it was a logical step in the right direction. That idea developed into the hot-line between the president of the United States and the soviet premier, following the Cuban Missile Crisis.

"Well, by the time I arrived in 1963, everybody had been talking about the study, and I had heard the rumors, seen the blips on radar, witnessed the commotions, and some of us occasionally even talked about the possibilities. But nothing really prepared me for what I started to read in the early morning hours one night in January 1964.

Aliens are Already Among Us

"It was about 2:00 a.m. and a relatively quiet night when the SHOC controller on duty went into the vault and came out with this huge document. `Take a look at this,' he said. The title was simply

"Assessment: An Evaluation of a Possible Military Threat to Allied Forces in Europe".

It was numbered, #3, stamped Cosmic Top Secret, had eight inches worth of appendices, dozens of photographs, and had been signed into the vault by German colonel Heinz Berger, SHOC's head of security. I quickly learned that it was based on two and a half years of research, was funded by NATO money, and that only 15 copies were published--in English, German, and French. Each one was numbered. All were classified and ordered to be kept under lock and key.

"Every time I got the chance, from then until I left, I would read a section or two in it. It was the most intriguing document I'd ever read. It was put together by military representatives of every NATO nation and also included contributions from some of the greatest scientific minds. These objects were violating all of our known laws of physics, and the study team had gone to Cambridge, Oxford, the Sorbonne, MIT, and other major universities for input on chemistry, physics, atmospheric physics, biology, history, psychology, and even theology, all of which were separate appendices.

"I read about theories on Einstein's sought-after unified-field theory, the high radiation at various landing sites, and UFO reports that dated back to

Aliens are Already Among Us

the Roman era and up to our own F105 pilots' sightings and encounters, and on and on. I had always been a skeptic, but this report, well...it concluded that this stuff was not science fiction.

"I read about contact encounters. One incident that had just happened in 1963 involved a landing on a Danish farm. According to the report, the farmer went aboard with the two little beings and two more human-looking men who spoke to him in Danish. The report included parts of his interrogation by government authorities and their conclusions that he was telling the truth.

In another incident, according to the reports, a craft landed on an Italian airfield and offered to take an Italian sergeant for a ride. He wet his pants--that's what it said--and was so scared, he didn't go.

"The appendix that really got to me was titled `Autopsies.' I saw pictures of a 30-meter disc that had crashed in Timmensdorfer, Germany, near the Baltic Sea in 1961. The British Army, according to the report, got there first and put up a perimeter. The craft had landed in very soft, loamy soil near the Russian border and so hadn't destructed, but one-third of it was buried in. We and the Russians, who also quickly showed up, had both tracked it.

"Inside, there were 12 small bodies, all dead. There were pictures of the bodies, which looked like the beings known as the `grays,' being laid out and then put on stretchers and loaded into jeeps, and autopsy photos, too. Some of the little grays appeared to not be a reproductive-capable species. The autopsy guys concluded, according to the

Aliens are Already Among Us

report, that it looked as if they had been cut out of a cookie cutter--clones with no alimentary tract. They did not ingest or process food as we know it, nor did it appear that they had any system for elimination.

"The craft itself was cut up like a pie into six pieces, put on lowboys and hauled off. Scuttlebutt was that it was given to the Americans and flown to Wright-Patterson Air Force base in Ohio. I looked at these pictures and couldn't believe it. My skin got cold and I thought, My God. I had never really believed we were all alone in the universe, but this was hard to swallow.

"The major conclusions in the NATO report blew me away. There were five:

1) The planet and human race had been the subject of a detailed survey of some kind by several different extraterrestrial civilizations, four of which they had identified visually. One race looked almost indistinguishable from us. Another resembled humans in height, stature, and structure, but with a very gray, pasty skin tone. The third race is now popularly known as the grays, and the fourth was described as reptilian, with vertical pupils and lizardlike skin.

2) These alien visitations had been going on for a very long time, at least 200 years--perhaps longer.

3) The extraterrestrials did not appear hostile since if that were their intent they would have already demonstrated their malevolence.

Aliens are Already Among Us

4) UFO appearances and quick disappearances as well as the flybys were demonstrations conducted on purpose to show us some of their capabilities.

5) A process or program of some sort seemed to be underway since flybys progressed to landings and eventually contact.

"I wanted so badly to copy this thing. I did take a photograph of the cover sheet, which wasn't in and of itself classified. But I didn't want to wind up in Fort Leavenworth. So instead I would go to the bathroom and take notes--surreptitiously, very carefully.

Aliens are Already Among Us

Aliens are Already Among Us

9.0 Other Interesting Information

In 2019 while I'm writing this book there have been some interesting reports in the news as follows:

> *April 24, 2019:*
>
> *The US Navy has updated its process for UFO sightings, with a new process to report sightings of 'unauthorized' and 'unidentified aircraft'.*
>
> *It comes after reports of sightings near military air space – although (of course) this doesn't actually mean the arrival of the little green men.*
>
> *Instead, the move is thought to be in response to sightings of unknown, highly advanced aircraft, some near sensitive military facilities, Politico reported.*
>
> *The US Navy said in a statement to Politico: 'There have been a number of reports of un authorized and/or unidentified aircraft entering various military-controlled ranges and designated air space in recent years.*
>
> *'For safety and security concerns, the Navy and the U.S. Air Force takes these reports very seriously and investigates each and every report.*
>
> *'As part of this effort, the Navy is updating and formalizing the process by which reports of any such suspected incursions can be made to the cognizant authorities.*

Aliens are Already Among Us

'A new message to the fleet that will detail the steps for reporting is in draft.'

Aliens and Volcanos

Although these pictures aren't part of a recent story I thought you might enjoy seeing it since some persons claim Aliens are able to hide near or inside of active volcanos. Looks like there is some truth to that claim from these pictures:

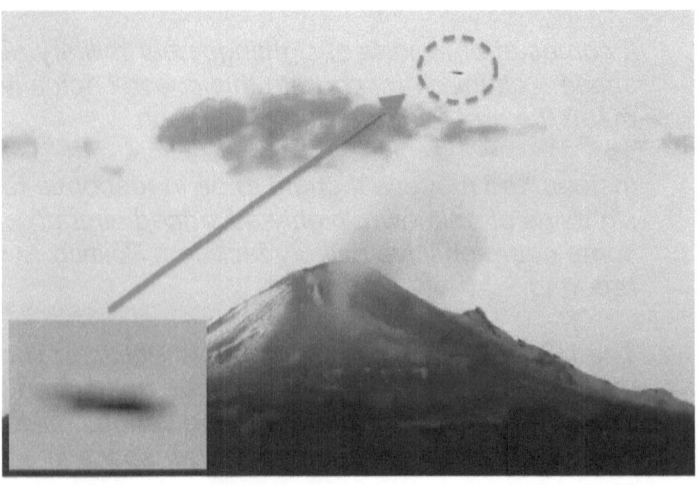

Aliens are Already Among Us

UFO appears to deliberately enter Mexico's Popocatepetl Volcano:

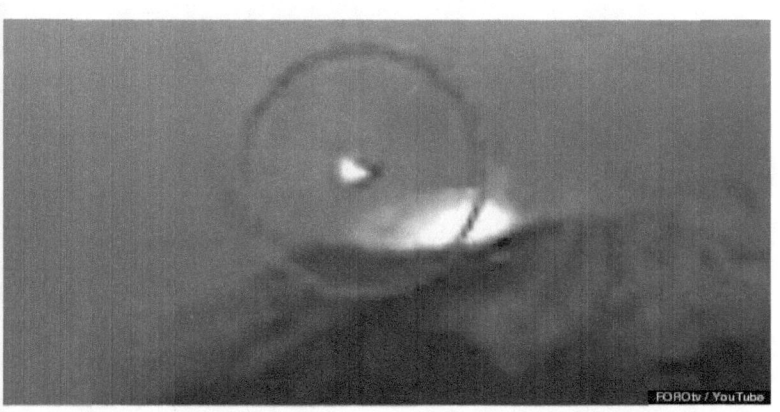

10.0 Alien Technology

a. Alien Invisibility

Any Aliens who could travel to Earth would by definition be thousands of years in advance of us technologically. What they could do would almost be akin to what we would perceive as magic.

Stealth technology for aircraft is only a few decades old here on Earth and it has already allowed us to make military jets almost invisible to enemy radars. What will this stealth technology evolve into over the next few hundred years?

There are also lots of experiments going on today to hide objects from visible light. Scientists are confident they can eventually make things invisible from light.

So, it is reasonable to assume that advanced Aliens can both hide their Spaceships and themselves from our eyes and sensors. There are some pictures in this book which indicate that sometimes alien invisibility fails and they can be seen or photographed.

b. Caret Alien Technology:

I have another book which is mainly focused on Alien Technology titled "Alien and Secret Technology-A Theory of the Hidden Truth".

This book is not about that technology which there is a good case really exists, but it should be mentioned briefly to understand the level of what their technology really is, and how it is probably thousands of years in advance of ours.

Aliens are Already Among Us

There are books claiming that we learned a lot of Alien technology from the crash of the Alien spaceship at Roswell New Mexico in 1947. (See the book "The Day After Roswell" By Col. Phillip J. Corso for the details.)

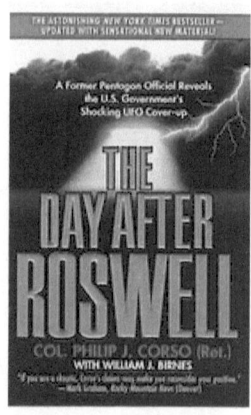

My beliefs in the incredible advances found in Alien technologies is founded on what was called the "Caret Project" and the information about their anti-gravity technology is shown below:

In 2007 a strange program came to light which is called the "Caret" program or "The Palo Alto Caret Laboratory" and which is supposed to have been an alien technology re-engineering program based at a civilian company in Silicon Valley in the 1980s. The work was performed on a classified contract for the U.S. military.

The story also includes eyewitnesses who saw a craft at different places in California which is supposed to be U.S. built and based on alien reverse engineered technology. A picture of the unmanned craft is below which has marking

Aliens are Already Among Us

later described to be similar to those from the Caret program:

What makes this story so fascinating are pages from supposed secret documents which provide a lot of details about the technology. The program was about sophisticated alien materials which could be programmed to have different physical properties including anti-gravity.

The idea is that these Meta materials have microscopic designs which allow the different properties to manifest purely by drawing the correct diagram. The technology was like writing software to make a program run, but instead of a program the materials would actually become what is programmed.

If true, this technology is obviously hundreds if not thousands of years in advance of ours because it means building Nano-machines at an atomic level and those machines having the ability to manipulate gravity too.

Imagine the ability to program physical properties and anti-gravity in an object just by setting up the correct program. Here is a picture of one of the programming diagrams:

PALO ALTO CARET LABORATORY   LINGUISTIC ANALYSIS PRIMER

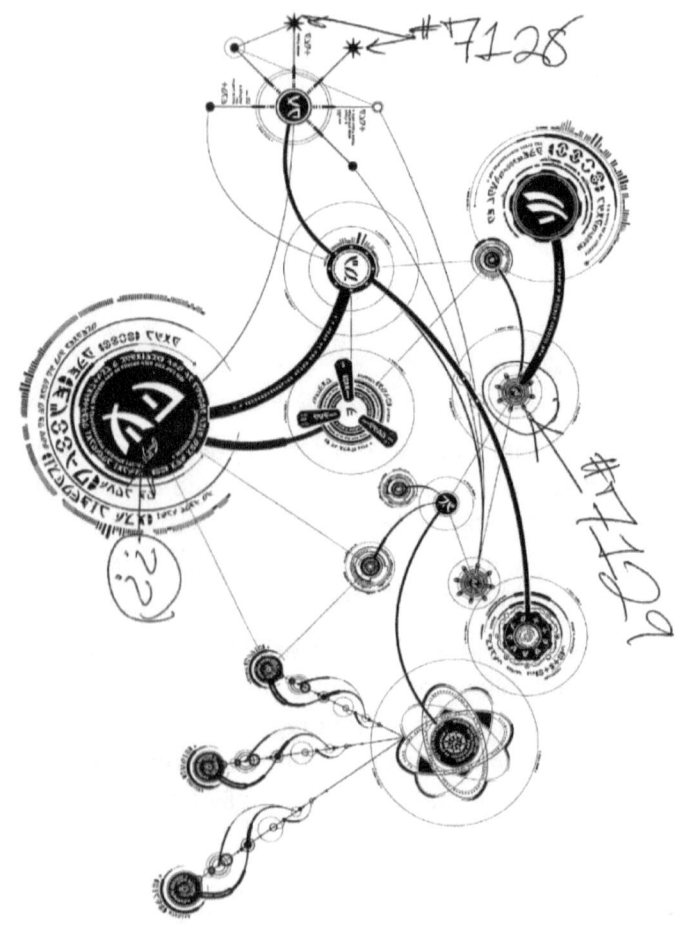

Figure 14.11
Full view of diagram D39-08-117c.

Another picture shows a Caret anti-gravity generator and two objects held in place by a designed anti-gravity field emanating from the round object in the center:

This program is so different from anything previously described in alien technology literature that I'm not sure what to make of it. The fact that unmanned craft were sighted numerous times by eyewitnesses makes me think this was a valid program.

Aliens are Already Among Us

Aliens are Already Among Us

11.0 Aliens on the Astral Plane

For those of you not familiar with the term "Astral Plane" it is the plane of existence where People's non-physical body is supposed to be able to travel to. It is achieved by a person having an "Out of Body" experience (OBE) which can be involuntary or voluntary.

OBEs have been experienced by many people and there are many books about those experiences as well as descriptions of the complexity of this non-physical spiritual environment.

I've had some experience with OBEs in my life although the experiences tend to be involuntary in my experiences.

Many persons who had OBEs also described meeting Aliens on the Astral Plane. It is a reasonable concept that

Aliens are Already Among Us

Aliens advanced thousands of years in advance of us are probably also more advanced spiritually and metaphysically. So what would be a better place to meet them than in a non-physical environment?

Here are some stories of persons claiming to have met Aliens in OBE experiences:

> *Adding to the overall weirdness is another recent account from a witness named Ian McLaren, who claims that he can also travel to other worlds through astral projection, even planets that we know nothing of. In this case he claims to have been to the ominous and mysterious planet Nibiru, a theoretical planet which some paranormal enthusiasts argue will at one point collide with Earth to cause a massive, extinction-event cataclysm. Yet according to this witness it seems like an alright place for the most part. He claims:*
>
> *At one stall I met two humanoid spirit guides who invited me to join them on a short journey to the Sirius Constellation which took about 10 minutes each way (there is no speed limit on astral travel) where we landed on planet Nibiru which was lit by three co-ordinated suns (Sirius A, B and C). When one sun set, the next in sequence rose so the planet was never in darkness. Nibirans are chosen from earth humans who have qualified and shown themselves worthy of immortality. As there is no such thing as death on Nibiru the only option for any spirits who are getting unduly bored with eternal life, is to return to earth for another cycle of reincarnations until they can again "work their passage" and can quality for transfer to a higher plane. Part of the test for life on Nibiru is to*

Aliens are Already Among Us

convince the assessment panel that you embrace the Quaker tradition of pacifist lifestyle as on Nibiru there are no Armed Forces. Death is not an option. There is no Police force as there is no crime. There are no medical facilities, nurses, doctors or hospitals as no one ever gets injured or falls sick. Everyone arrives at Nibiru with a comprehensive education so no schools, colleges or University with their associated stresses. On arrival my two Spirit Guides logged me in to the Visitor's Book and we started our tour. Clothes are optional as it is a very comfortable climate.

We walked forward along a path and into a fresh water lake. Having fully functioning gills/lungs there was no risk of drowning. I stroked a fish of the style of a dolphin and lay on its back while it took me on a tour of the lake. I noticed that aircraft were solar powered with having limitless light. Speedboats derived their power from the water and the use of dis-similar metals. All parachutists landed safely so no broken ankles and they were able to walk away from every landing. All food options were vegetarian. It is my choice of planet if I maintain a good record for the remainder of this incarnation. I've had quite enough of the way things are going on this plane even if it is relatively civilized compared to the lower, darker planes.

Aliens are Already Among Us

Aliens are Already Among Us

12.0 How to contact Aliens

If there are Aliens living on or visiting the Earth, then maybe there is a way to contact them.

I've often thought that there might be a way and so I turned to the Internet. If you are a real Alien and you want to avoid publicity and keep your identity secret, then this would be one of the best places to go I thought.

In searching the Internet I found that I wasn't the first person to do so, and there are even some interesting stories of claims by beings on the Internet to be visitors on Earth. Here is one of the most interesting stories:

> *Taylor Genovese tells this story—*
>
> *In April 2009, I was browsing the Something Awful forums when a particular thread caught my attention:*
>
> **I am an alien, and I would like to talk to you about it.**
>
> *The thread began with a simple greeting: "Hello. As I stated in the title, I am an alien. More specifically, I am from a planet two arms away from this arm of the galaxy."*
>
> *The self-proclaimed alien went on to explain that we should address him as Mr. Boone, a name he had taken from the first person he had encountered after landing on Earth. He said he had been visiting our world since 2007 and was doing so recreationally. He had landed in the United Kingdom and chose Something Awful because it was the "Internet area comprised of the largest mix of differing and open-*

minded intelligences." He continued by describing his planet—called Ulath—as very similar to Earth, except that it is slightly larger, 40% water and 60% land, and is much colder. His species—known as melrins—were similar to humans in that they had two arms, legs, ears and eyes as well as a torso, head and mouth. However, they lacked noses, instead smelling with their mouths, and had thicker keratinous skin on their backs that developed due to the large hailstones that fell to the surface during storms.

I was hooked. And I was not the only one. Soon, the thread exploded with users asking questions ranging from galactic politics (a sort of anarcho-socialist utopia comprised of a network of cooperative decentralized planets) to how his race pooped (the same mechanism as humans—but more liquid than solid due to a vegan diet). In all he answered 531 questions over the course of three months, along with several lengthy posts describing his world, his life, and how these life-forms capable of star travel interact with one another.

Aliens are Already Among Us

An image posted by Mr. Boone indicating the region where his home planet of Ulath is located within our galaxy.

What really impressed me, however, was how the author—or Mr. Boone—approached the little details in his story, particularly the linguistic differences between his language and English.

"I am here on a [recreational] and exploratory visit, as is one of my personal [hobbies]. (I use square parenthesis to denote where I have used a word or phrase that is not entirely accurate, but the closest approximation in English)...The phonetic spelling of my name is approximately awr-bt-n[yi]-ah. The 'yi' is in square parenthesis as it is not quite representative of the vowel sound my language has that yours does not. While on Earth, Mr. Boone is

Aliens are Already Among Us

my preferred name. I am 58 years old, identify myself as an unskilled [scientist], and have travelled to 7 [lonely] planets and nearly all planets of the [network], of which there are more than 20. My excursions to [lonely] planets last between six months and three years, and I produce no in-depth scientific reports when I return from them. I intend to stay the full three years and perhaps even longer on Earth, as your planet is in extreme turbulence and I find it engrossing."

This clever use of brackets helped add an additional layer of depth to his story and allowed for quick transliteration on the fast-paced medium of an internet forum. He skirted around more complicated questions about astrophysics, mathematics and demands of proof that he was an alien by blaming humankind's current violent attitudes. According to Mr. Boone, before having the ability to travel the stars, humankind must eliminate violence or the network would intervene.

"If I were to prove that I were an alien, your government would come looking for me and I would be forced to go home. After that, they would know for sure that there was extra-terrestrial life, and so would everyone. It would shake up your society violently enough to change where you are going, and you might develop things before you normally would have, and maybe get out into the galaxy before you stop blowing shit up. That wouldn't do, so no. No proof."

Although the network that Mr. Boone describes sounds great on the surface— no war, total cooperation, and a society built for the betterment of

Aliens are Already Among Us

everyone regardless of any differences—it also had oppressive elements, such as the total eradication of religion and belief as well as carrying out eugenic genocide.

"You are talking about eugenics, and yes, it happened. In a big way, and not in a way you would consider ethical...The eugenics happened shortly after war was all but eradicated, and were nothing to do with skin colour. They were more about what was, at the time, considered 'disease'. Genetic disorders, and some people were prevented from breeding in case they spread certain viewpoints."

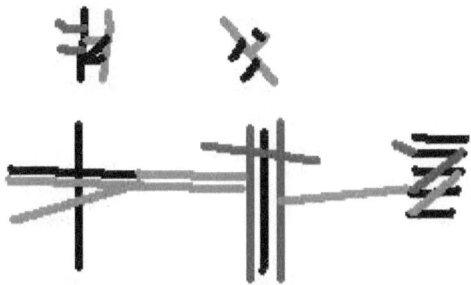

An example of melrin written language courtesy of Mr. Boone. They are his name (top left), his purpose—he described as "retail-management tourist" (top center) —and the name he used when submitting entries into the network databases (bottom).

He also described a system of "civilized development" that—much like the Kardashev scale—had an uneasy similarity to inaccurate, colonialist anthropological ideas of primitivism, and Eurocentric ideas about how a "civilization" develops (i.e. tribes to city-states to nations, etc.).

Aliens are Already Among Us

Included in this scale is vernacular that would make any anthropologist cringe (i.e. using the term 'savage'). Mr. Boone claimed that "all planets we've discovered so far follow this general pattern."

- *Class 1* - Savagery
- *Class 2* - Calculated Savagery
- *Class 3* - Educated Savagery
- *Class 4* - Purposeful Savagery
- *Class 5* - Restructure
- *Class 6* - Post-Restructure

A condensed explanation provided by Mr. Boone:

• Class 1 societies are equivalent to the prehistoric.
• Class 2 develop class hierarchy, slavery and war.
• Class 3 is defined by warfare and battling between religion and science.
• Class 4 has significant atrocities that bring rebellion by the people.
• Class 5 develops widespread space travel and abolishes trade.
• Class 6 is—essentially—utopia.

On July 31, 2009, Mr. Boone decided to leave Earth to a chorus of gratitude and wishing of luck from the community. He signed off by telling eastern UK residents to look at the light side of the moon because they may see a "shooting star" that would be his ship. He left our planet with an: "Adios, aloha, auf weidersehn, goodbye!"

Aliens are Already Among Us

In 2012, Mr. Boone resurfaced by sending an email blast to all those who had previously emailed him. Another thread appeared on Something Awful announcing: **Mr. Boone is back!** *Supporters and critics began to emerge from the recesses of the community. Mr. Boone posted that he may create a separate thread to update the community on his travels and allow for more questions and answers. He also indicated that he had "friends in tow" this time.*

That thread never materialized and it was the last anyone has heard of Mr. Boone. However, user Leovinus revealed that he was the one who—in 2009—set Mr. Boone up with a laptop computer and a Something Awful account. Whether or not this is the truth—or a continuation of Mr. Boone's fictional legacy—is irrelevant. What is relevant is the realization that online community-based speculative fiction is an on-going process rather than a traditional project with a beginning and an end. It exists within a collective imagination that can be fed-into and drawn-from symbiotically by the entire community.

Aliens are Already Among Us

Aliens are Already Among Us

13.0 Summary

So what do you the reader think after reading this book?

Have I made the case to you for Aliens having visited our Earth now or in the past?

To me the odds from the Drake Equation strongly favor a large number of advanced Alien civilizations in our Galaxy.

And there are too many reports of Aliens in history or currently on Earth to account for. There must be some truth to them being here instead of everything being fabrications by wild eyed people or conspiracy theorists.

Admittedly, there are a lot of Kooks out there—And I met a number of them at a recent UFO conference—"Contact in the Desert" in 2017. Many of the speakers were real nut cases or didn't present any evidence of what they were talking about, but I saw at least one person—Dr. Stephen Greer who made a logical case for Aliens visiting our planet.

I guess my conclusions draw from the huge number of stars and planets which exist in the cosmos, and therefore the number of intelligent civilizations must be huge.

Since it is so huge, many of them must have visited our planet from eons past until now. And they all have the ability to hide themselves from us. Both their spaceships and the Aliens themselves can become invisible due to their advanced technology.

Most persons' refusal to admit that intelligent Alien life exists are only holding onto age old prejudices like our ancestors who thought that the sun and planets revolved around the Earth.

Aliens are Already Among Us

Aliens are Already Among Us

Bibliography

1. https://www.sacred-texts.com/ufo/military.htm. *https://www.sacred-texts.com.* [Online] 1997.

2. https://mysteriousuniverse.org/2019/02/truly-bizarre-accounts-of-astral-travel-to-other-worlds/. *https://mysteriousuniverse.org.* [Online] 2019.

3. https://medium.com/space-anthropology/aliens-on-the-internet-60f72332ec81. *Aliens on the Internet.* [Online] 2015.

4. https://shape.nato.int/resources/21/alleged%20shape%20assessment%20of%20ufos.pdf. *Shape Assessment of Aliens.* [Online] 1964.

5. Greer, Dr. Stephan. *Disclosure.* 2001.

6. Cook, Nick. *The Hunt for Zero Point.* 2007.

7. https://listverse.com/2016/06/13/10-alleged-underwater-alien-ufo-bases/. *10 Alleged Underwater Alien Bases.* [Online] 2016.

8. Corso, Col. Phillip J. *https://en.wikipedia.org/wiki/The_Day_After_Roswell.* 1997.

9. https://proofofalien.com/top-10-underground-alien-bases-on-earth/. *Top 10 Underground Alien Bases on Earth.* [Online]

10. Hall, Charles James. *MIllenial Hospitality.* 2007.

Aliens are Already Among Us

Aliens are Already Among Us

Index

100 million planets, 5
age of our Universe, 4
alien craft in flight in 2004, 1
alien who thought he was invisible, 6
Aliens and Volcanos, 78
Aliens Underwater, 53
Ancient Alien Base At Lake Titicaca, 62
Ancient grooved spheres, 9
Anunnaki, 10
Area 51 underground alien base, 52
CARET Programming Diagram, 84
Cave entrance of Mt. Lassen, 50
Caves in Colorado, 48
Dogon Tribe, 15
Dulce Base in New Mexico taken from Google earth, 51
Dulce, New Mexico, 40
EBE #1: The UNTOLD STORY OF ITS LIFE, 26
Ezekiel and the flying machine, 13
Flight Corridor To Underwater Base North Island, New Zealand, 58
fresco entitled "The Crucifixion, 14
Full Drake Equation, 3
Increased UFO Sightings At Lake Erie Cleveland, Ohio, 65
Indian Springs, California, 37
invisible alien, 6
Lake Baikal—Scene Of An Underwater Battle With Aliens, 66
Legends of the Annunaki, 10
Mt. Shasta underground base, 49
NATO Top Secret Study, 69
Nommos, 16
Out of Body, 87
Pacific Ocean 'Humming' Is Actually Marine Life Breaking Wind, 59
Paintings on Stone walls, 12
possible underground tunnels, 41
Puritans in New England, 17
Roswell Incident, 25
Roswell New Mexico, 25
Serpo—Missions to Other Planets, 31
Sightings in the Nineteenth Century, 21
Strange Lights And Sounds In Puerto Rico, 60
The Day After Roswell, 82
The Drake equation, 3
The Dulce Base, 40
The Palo Alto Caret Laboratory, 82
The Serpo project, 31
UFO Base Under Lake Ontario, 56

Aliens are Already Among Us

Underground alien bases in America, 47
Underground alien bases in Canada, 44
Underground alien bases in Europe, 43
Underground alien bases in Russia, 46
Underground Base of the iron mines in Newfoundland, 45
Underwater Alien Base At Guantanamo Bay, 63
Underwater Base Off Puffin Island Wales, 55
Underwater Base Off The Coast Of Malibu, 53
Zeta Reticuli, 35

www.ingramcontent.com/pod-product-compliance
Lightning Source LLC
Chambersburg PA
CBHW021833170526
45157CB00007B/2794